BLOOD!

Not Just a
VAMPIRE DRINK

I DON'T VANT TO SUCK YOUR BLOOD
SMOOTHIE SHOP

By Stacy McAnulty • illustrated by Shawna J. C. Tenney

GODWINBOOKS

Henry Holt and Company • New York

Good evening. A pint
of your best blood,
type O-positive.

For Lily, who donates blood and saves lives
—S. M.

To my BLOOD sisters, Bethany and Diana
—S. J. C. T.

Henry Holt and Company, *Publishers since 1866*
Henry Holt® is a registered trademark of Macmillan Publishing Group, LLC.
120 Broadway, New York, NY 10271 · mackids.com

Our books may be purchased in bulk for promotional, educational, or business use.
Please contact your local bookseller or the Macmillan Corporate and Premium Sales Department
at (800) 221-7945 ext. 5442 or by email at MacmillanSpecialMarkets@macmillan.com.

Library of Congress Cataloging-in-Publication Data is available.

First edition, 2022
Book design by Mina Chung and John Daly
Painted digitally in Photoshop with gouache, pastel, and splatter brushes
and in Procreate with chalk, gouache, and watercolor brushes.
Printed in China by RR Donnelley Asia Printing Solutions Ltd.,
Dongguan City, Guangdong Province

ISBN 978-1-250-30405-6 (hardcover)

1 3 5 7 9 10 8 6 4 2

Sorry, Count. We don't serve blood here. It's too important to humans for us just to drink it.

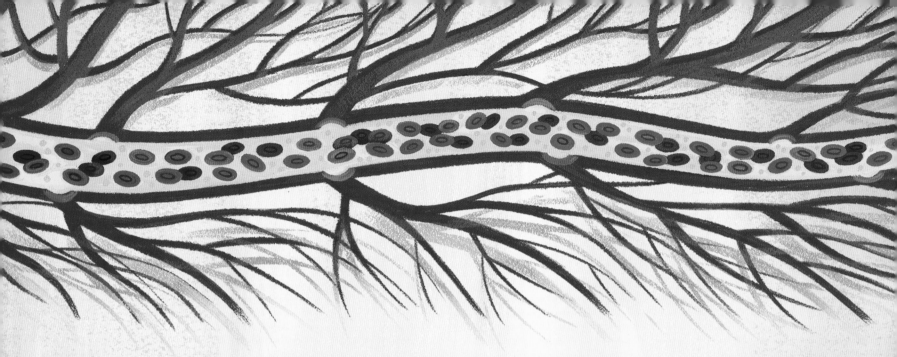

Blood doesn't belong in my blenders.
It belongs in the cardiovascular systems—the body's highway for delivering
oxygen and nutrients to organs and hauling away the waste.

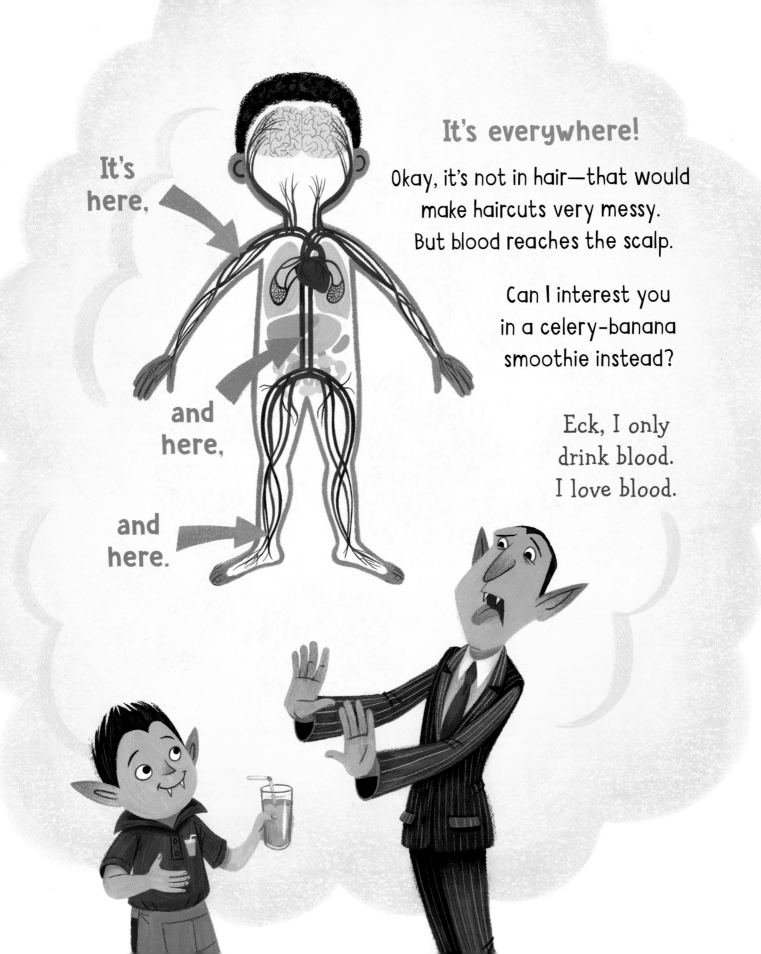

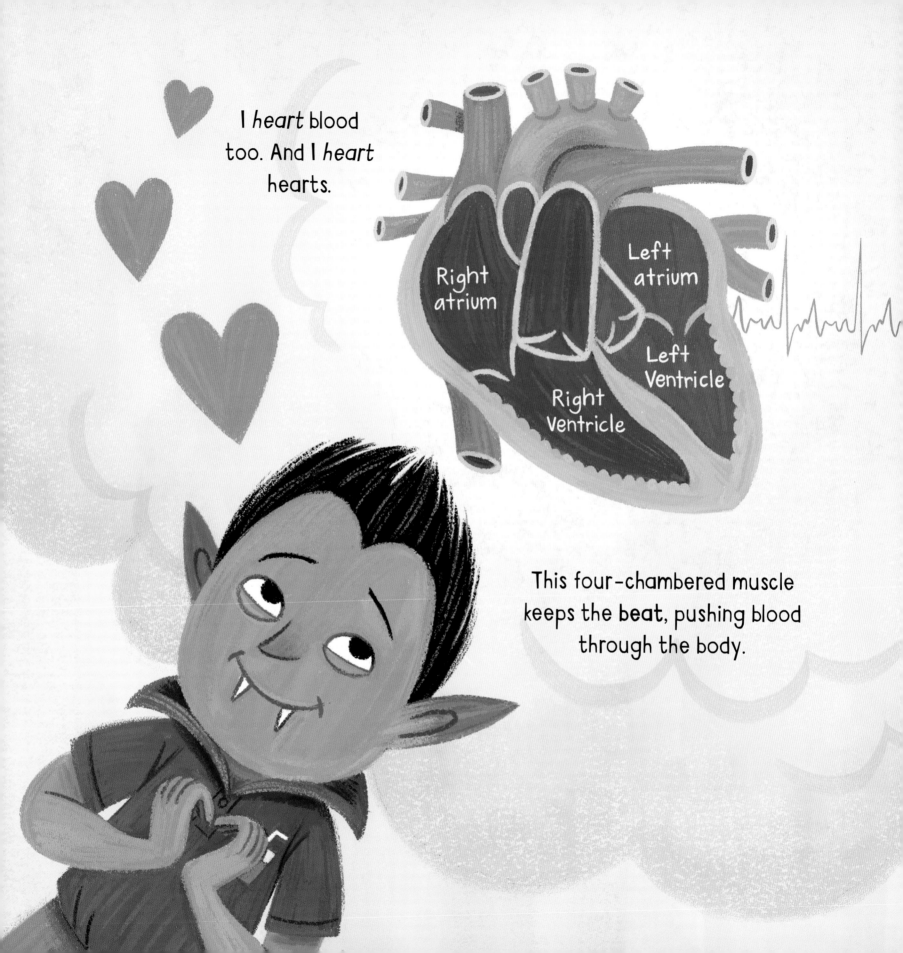

I *heart* blood too. And I *heart* hearts.

Right atrium

Left atrium

Left Ventricle

Right Ventricle

This four-chambered muscle keeps the **beat**, pushing blood through the body.

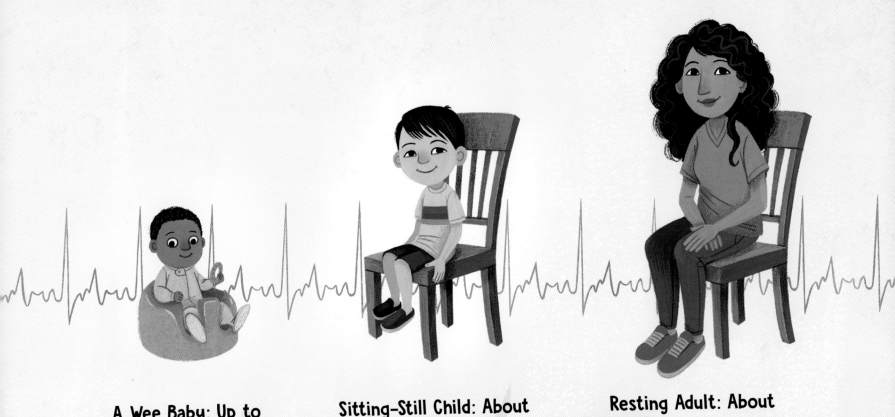

A Wee Baby: Up to
160 Beats per Minute

Sitting-Still Child: About
80 Beats per Minute

Resting Adult: About
60 Beats per Minute

From the heart, blood flows . . .
Through the arteries
(tubes with muscles),

then to the arterioles (branches of tubes),

then to the capillaries (teeny-tiny tubes)
that can get to teeny-tiny places.

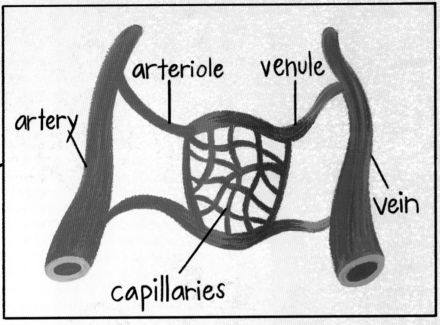

artery

arteriole venule

vein

capillaries

Now the blood must get back to the heart . . .
From the capillaries to the venules
(more branches of tubes),

then through the veins
(thinner tubes without muscles),

and, finally, home
to the heart.

Vhen does it flow
into my mouth?

How about a pineapple-
fig smoothie?

Vhat is in it?
Any red blood cells?

Well, no, it has pineapple and figs.
And blood has more than red blood cells.

Blood Ingredients List

Red Blood Cells

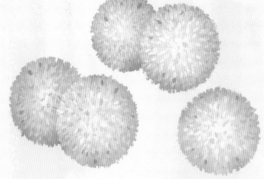

White Blood Cells

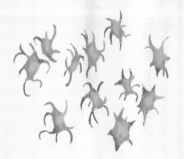

Platelets

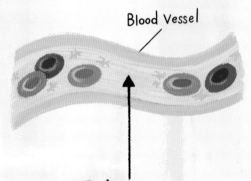

Blood Vessel

Plasma

The red blood cells
carry gases.

They pick up oxygen from
the lungs and carry it to the
muscles, the brain, the stomach,
everywhere.

And then they take away the
trash—the carbon dioxide.

When filled with oxygen, red blood cells are bright red— thanks to hemoglobin.

Without oxygen, the cells are a dull red.

From artery

Red blood cell carrying oxygen

Capillary

Red blood cell with no oxygen

To vein

Oxygen

To body's cells

Carbon dioxide

From body's cells

Red is my favorite color. After black, of course.

Most animals have beautiful red blood, but not all.

Spiders and horseshoe crabs have blue blood.

Blech!

Ocellated icefish blood is clear.

Ick!

Leeches and some worms have green blood.

Gross!

I'd rather lick a Band-Aid.

If red blood cells
are too important,
give me a glass of
vhite blood cells.

The vhite—er, um, white blood cells—
are just as important.
They are mighty warriors,
fighting off disease and infections.

Vhere? Vhere is this delicious child with the fountain of blood?

You're silly, Count.

Vhat I am is THIRSTY!
No red. No vhite. No platelets.
Vhat is left?

Plasma!

Human blood is mostly plasma—the red cells, white cells, and platelets all swim in it—and plasma is mostly water.

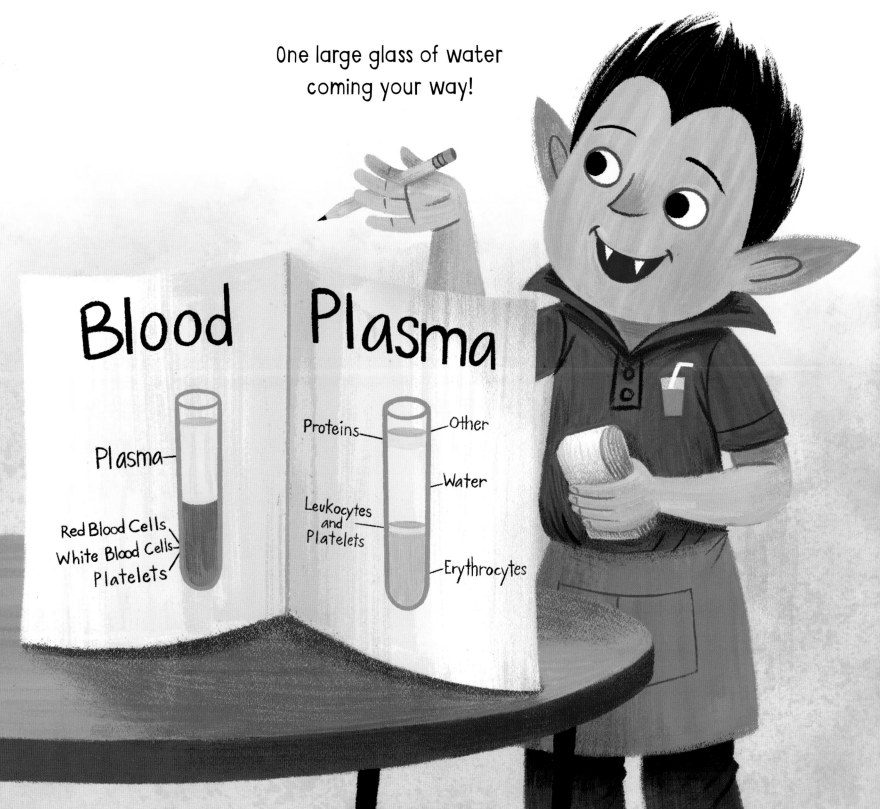

I don't vant vater. I don't vant a pumpkin-rhubarb smoothie.

I VANT TO DRINK BLOOD! JUST ONE GLASS, PLEASE.

No can do, Count.
While grown-up humans have over a gallon of blood,
we can't just down it for dinner.

16 cups = 1 gallon

Though people can
donate blood.

Yes. Yes. They can donate blood to me!

Nope. Nope.

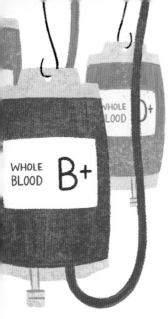

They can donate blood to help other humans, not to feed a thirsty vampire.

They can do this because the body makes new blood all the time.

Basically, people are blood factories.

I know! I know!
You're killing me, kid.

Blood cells are created in the bone marrow—
the spongy, soft center of human bones.

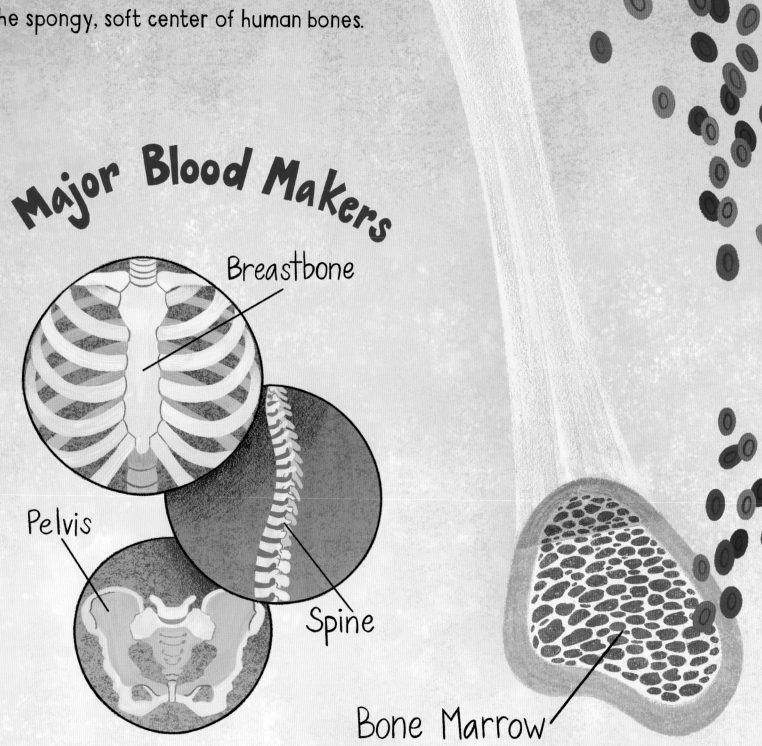

Major Blood Makers

Breastbone

Pelvis

Spine

Bone Marrow

Vait just a second.
You're a vampire,
and you don't drink blood?

Nope. It has too many extra
ingredients. Like traces of lead,
zinc, iron, and even gold.

I've been drinking blood for
397 years. It is my favorite.
I've never even tried
anything else.

Seriously?

I'm very serious. I do not joke.

Then I've got something for you to sink your fangs into.

A chocolate milkshake!

Blah. It's the wrong color.

Try it.

It's the wrong temperature.

TRY IT!!

I think you've
earned this shirt.

No longer
A
SUCKER!

DEAR BLOOD-FILLED READER,

Blood is such an important part of the human body—it delivers nutrients, hauls away waste, fights infections, regulates temperature, and more! It's easy to imagine why vampires love the red stuff. And while we can't share our blood with funny, fanged monsters, healthy adults can donate blood because the body is always making more.

Donated blood is used to help people during surgery, after an accident, or with some illnesses. It's an incredible fluid that can save lives, making it seem almost magical. According to the American Red Cross, someone in the U.S. needs a blood transfusion every two seconds. And all of that blood comes from donors, not factories or laboratories (or smoothie shops).

So, if you ever meet a vampire (you won't, they're not real), tell them that blood is a terrible beverage but an awesome part of your body.

Sincerely,
STACY
Author and blood-making machine (Type A+)

SOME BLOOD FACTS

Humans have different blood types:

A+, A-   B+, B-   AB+, AB-   O+, O-

And O+ is the most common. Scientists are still trying to understand why our species developed blood types, but they do know you inherit your blood type from your birth parents.

About 10 percent of a grown-up's weight is blood. For an average man, that's about 9.6 to 12 pints (or 1.2 to 1.5 gallons).

A healthy adult can donate 1 pint of blood every 56 days with no negative effects. And they get cookies and juice when they're done.

Blood is precious—so precious, in fact, that there are small amounts of gold in it! If we could mine that gold, we'd get about 0.2 milligrams, which is about the weight of a mosquito.

Red blood cells have a lifespan of 120 days, while platelets only have a 10-day span. Depending on the type of white blood cell, it can live for hours or years.

Hemophobia is the extreme fear of blood. Someone suffering from this would be bothered by the sight of blood and might have difficulty with needles and shots. They also might not enjoy this book.

Vampires are fictional, but a few real-life critters *do* drink blood: mosquitoes, leeches, fleas, ticks, and some species of bats. Humans have enough blood that these animals won't drink a person dry; however, these bloodsuckers can spread disease.

Legend tells us that vampires' favorite things are blood, coffins, and castles. They hate sunlight, garlic, and wooden stakes.

SOURCES

Bailey, Regina. "Top 12 Interesting Facts About Blood." ThoughtCo, June 21, 2019, www.thoughtco.com/facts-about-blood-373355.

"Blood Basics." American Society of Hematology, www.hematology.org/education/patients/blood-basics.

"Blood Types." American Red Cross, www.redcrossblood.org/donate-blood/blood-types.html.

Conley, Lockard C., and Robert S. Schwartz. "Blood." *Encyclopedia Britannica*, Updated October 24, 2020, www.britannica.com/science/blood-biochemistry.

Dallas, Mary Elizabeth. "10 Amazing Facts About Your Blood Vessels: Everyday Health." EverydayHealth.com, July 20, 2015, www.everydayhealth.com/news/10-amazing-facts-about-your-blood-vessels/.

"Facts About Blood." Johns Hopkins Medicine, www.hopkinsmedicine.org/health/wellness-and-prevention/facts-about-blood.

"How Your Body Replaces Blood." NHS Blood and Transplant, www.blood.co.uk/the-donation-process/after-your-donation/how-your-body-replaces-blood/.

"Importance of the Blood Supply." American Red Cross, www.redcrossblood.org/donate-blood/how-to-donate/how-blood-donations-help/blood-needs-blood-supply.html.

"The Chemistry of the Colours of Blood." Compound Interest, October 30, 2015, www.compoundchem.com/2014/10/28/coloursofblood/.